Charles Barnard

My Handkerchief Garden

Charles Barnard

My Handkerchief Garden

ISBN/EAN: 9783337375270

Printed in Europe, USA, Canada, Australia, Japan

Cover: Foto ©berggeist007 / pixelio.de

More available books at **www.hansebooks.com**

MY
HANDKERCHIEF GARDEN.

CHAPTER I.

ꜧOW IT BEGAN.

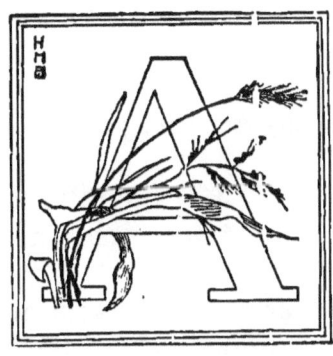

ALONG the edge of the Sound, from Stamford to New York, we had looked everywhere in the hope that we might find a small house, a little garden, and a low rent. These things seldom grow together. Houses with no land, land enough with big houses, and both land and houses in plenty at high rents. At last it was found; a six-room house with a mere handkerchief of a garden, measuring about one-thirtieth of an acre, or about as big as a city back yard. The soil was a wet, heavy clay, full of stones, and shaded by a number of tall trees growing on the next lot. In March, 1887, we moved to the place, and on the twenty-first we paid twenty-five cents for one ounce of Tennis Ball Lettuce seed. So it was the scrap of a garden began, and thereon does hang the more or less learned remarks that make this book. There are people so constituted that they cannot see anything remarkable in a paper of seeds. A seed is potential wealth—bran new wealth that does not exist, but waits the partnership of nature and the gardener. Seeds are about the cheapest thing in the world. At wholesale a cent will buy a hundred seeds of lettuce. An acre of ground, if managed by a man who knows his trade, will produce in one season 40,000 heads of lettuce. New York will calmly eat every head at three cents each and cry for more. You would probably pay at the store five cents a head or $2,000 for the lot.

Oh! Figures can be made to say anything.

Think so?

All the same, you and I and the rest of the folks do pay $2,000 to somebody for that yield of lettuce many times over every spring.

It's quite true the actual grower may not get it all. He seldom does and thereon might be written a tale of woe that would move the world to tears had it not, poor world, been listening to the gruesome story for about seven thousand years. There are problems in social economies so old that they have lost the power of speech. This is one of them, and it was in our handkerchief garden we dug up a great truth that may help to solve this very problem. It was a dusty old truth and smelled of the earth, yet, by decking it out in a few sprigs of pungent parsley, and framing it with the enticing lettuce, the persuasive pea, and the inspiring cauliflower, I hope to set it forth as a dish worthy the intelligent reader's grateful digestion.

A garden is a queer place. You can dig up facts and greens with the same hoe—provided you know a fact when you see it. Therefore, it happens that in presenting this dish of greens I may add sundry facts for dressing, the same facts being duly dug up in the same handkerchief garden. Persons of a romantic and expansive frame of mind have written about gardens without regard to the facts growing therein. One particularly aggravating person even wrote a tearful account of one sad summer in a garden, and the world has been delighted to read it many times over. My handkerchief garden is on a plane of more solemn import, and its great moral lessons are not addressed to persons of a light and sportive nature. They appeal rather to those finer instincts of the heart that cluster round the dinner table and the green-grocer's bill.

A pinch of seed in a paper bag is about as useless a thing as you can find. The seed must be joined to three great facts in nature—heat, moisture and light, if new wealth is to reward our labors. Lettuce seed will sprout in a temperature ranging from $60°$ to $70°$ by day, and not less than $40°$ by night.* Our room had a west window and was warmed by a small stove. Here were the elements of horticultural success. There was no gas in the room. This is most important—for in my experience it is difficult, almost impossible, to raise plants of any kind where gas is burned. A single gas-jet will spoil the air as fast as six men. Plants must have pure air, and as they can not go out for a walk every time they have a headache, they give up in despair in a room where a gas-lamp has ruined the air for breathing purposes. The moral of all this is just here; if plants will not thrive in bad air, how can we? We do not. There is, it is true, no authenticated record of a death by reason of a gas-light, yet it is also true that we are hurt in greater

---

* All temperature records here are by Fahrenheit's scale.

or less degree by every lamp that burns in our houses. The real moral is—ventilate your rooms.

To secure the union of the three facts, heat, light, and moisture, and bring them to our lettuce seed, we procured a six-inch flower-pot, worth, perhaps, ten cents at retail. By dint of digging in the garden I managed to get a few lumps of half frozen soil, and put them to dry on the floor by the fire. In a few hours the soil was soft enough to break up into dry mould in the hand, and I filled the pot nearly full, scattered the lettuce seed on top, sifted more soil over it through the fingers, and gently pressed it down firm. A sprinkle of water shaken over the soil by hand, wet the seed, and then the pot was placed in a warm corner behind the stove, and covered with an old newspaper. Two factors were thus provided, heat and moisture. The third could be added afterwards, as soon as the seeds began to stir with young life. For three days the flower-pot garden was examined night and morning, and, if the soil was dry, a little more water was added. On the fourth day the surface was broken by tender young things just poking their green fingers up to reach the light and air. The pot was at once placed in the window, and there it stayed about six weeks, and was then completely filled with young lettuce plants about three inches high, and hanging over the sides of the pot in a luxuriant, pale-green mass. So it was my handkerchief garden began in my study window.

CHAPTER II.

WHAT WAS DONE WITH IT.

WHEN the snows of March melted away, the garden came into sight. The former tenant had apparently regarded the garden as the proper place to deposit the waste of a generation. Bones, clam-shells, rejected shoes and cans, were plentiful. Added to this, it had not been dug over since the last crop, and corn-stubble covered half the space. The carpenters at work on the house had tramped the soil down hard, and in a corner under the trees were the remains of countless weeds nipped by last year's frost.

A very slight examination showed that the soil had one great merit. It was strong. A mass of rocks, weathered by the storms of a hundred years, and grey with moss, had sent down their fertilizing dust, and the tall trees had every year carpeted the place with their leaves. There had also been hens and pigs on the place, and these, too, had done what they could to contribute to the future crops. Rumor also had it that in the previous year it had been heavily manured, and had borne a large crop of corn and beans. Here was the problem. The place measured about eighty feet on one street, and seventy-five on another. The house stood just south of the center, near the street. A rocky cliff behind the house, while very picturesque, was, of course, valueless for any purpose, being too steep for a foothold, and too bare to produce anything save mosses and lichens. What could we do with it. The most simple way to treat the place would be to sow grass seed over all the ground, and keep it in grass. Merely to let it run up to grass would be cheap but ugly. If in grass at all, it must be kept as a lawn. A lawn would certainly look well, save all care and expense, except the weekly clipping to keep the grass in order. That means a lawn-mower, costing ten dollars. It means labor in pushing it over the grass, not less than fifteen times every summer. It is doubtful if this could be done for less than fifteen dollars. Some of my neighbors tell me it costs $1.50 a week for five months each year to keep a small lawn in order. The cost on this place, including the mower, would not be less than twenty-five dollars the first year. The cost of preparing the ground and sowing the seed would not be less than four dollars more, and each spring, fertilizer to the value of two dollars would be required. It began to look as if the cheapest thing that could be done would be pretty expensive. Of course, the place could be left to take care of itself, but this would be morally wrong. There were gardens on every side kept free from weeds at a greater or less expenditure of time, labor and money. To suffer weeds to bloom and scatter their seeds over these gardens, and thus to injure the neighbors' property, would be inexcusable. No man has a right to propagate weeds near any cultivated land. It is simply unjust to permit weeds of any kind to grow on your land while others are trying to keep them out of their land. A lawn is, therefore, a moral measure, as it checks the growth of weeds, and by its beauty enhances the value of the estate, and of all those near it.

That settled the matter. Something must be done with the ground. It must be either laid down to grass, or cultivated as a garden. The chief cost of a

lawn is the labor. There was my own labor. Could
I not push the lawn-mower myself? Many of the
gentlemen near by did so. Why could I not do like-
wise? No reason whatever why I might not in this
way save part of the expense of a lawn.

Now a lawn-mower is very well in its way. It's not
very hard work to use it, and it keeps a man out in the
air and sunshine. The chief objection is that it is not
work enough. It pays to work out of doors. For
every man who works a part of the day in the house
there should be several hours devoted to exercise in
the open air. A garden is a sanitary measure. It
takes you out on the sweet, healthful ground. A garden is a good place to bury headaches. That settled
the matter, and I decided to use all the available land
for a flower and kitchen garden. There were two
other reasons, beside the sanitary advantage, for
having a garden. In suburban towns and villages
the rent is for the house, and the lot of land on which
it stands is practically thrown in free. It costs no
more to have the house without the land than with it,
for as soon as the land becomes too valuable, the
houses cover all the land as in a city. If the land is
used for a garden it will make a solid financial return,
while a lawn pays nothing beyond the doubtful value
of looking pretty from the road and the Christian
grace of doing as you would be done by in the matter
of weeds. All this had been settled when the lettuce
seed was bought, and on the seventh of May, 1887, I
put spade in the new venture.

It wasn't really a spade, for a digging fork is better.
On that seventh of May I bought a digging fork, hoe
and steel rake at a total expenditure of $1.38, and the
handkerchief garden began. I had previously bought
for $1 60 ten plants of the "Jessie" strawberry, and
they had been kept in a friend's garden while absent

from home during a part of April. By May seventh all were dead save five, and the first work done was to fork up a little spot in the garden and set out those five plants. On the same afternoon the pot of young lettuce plants was brought out to the ground and a place, four feet square, was forked up and made smooth. On this little bed was set out, about five inches apart, a few dozen lettuce plants. There were some left which were given away to a neighbor—the first crop for the season.

The soil proved to be very tough and stiff, and one Italian man spent one day in trying to spade it over and nearly perished in the attempt. After that I did all the work myself, forking up the ground in little beds as it was wanted. This labor, with the tools and more seeds, brought the expense of the garden on the first of June up to $5.88. On the thirteenth of June the first lettuce and radishes were placed on my table, and the garden was credited with the first return, five cents for a fine head of lettuce and five cents for a bunch of breakfast radish.

The price of a fair average head of lettuce in the village store on that day was five cents. I had had been paying that sum every day for two or three weeks, and often paid more. The day I picked that head of lettuce I saved five cents on the bill at the store. It was perfectly fair and right to credit the garden with the retail price of lettuce for that day.

My labor? Oh! yes! It cost labor to raise it, cost the seed and the flower pot, and all the little procession of odd minutes spent in caring for the crop. These were worth money, if my time was worth anything at all. My time is worth something for about five hours out of the twenty-four. The time spent in caring for the lettuce plants was simply unavailable or "off time" of no value, except as a time for exercise.

Exercise then must be, and was it not better to raise lettuce for my table than to trundle an unprofitable lawn-mower, or walk the streets in idleness. Beside this, every hour spent in the garden was a sanitary gain and therefore a commercial gain that could not be expressed in money. I am certain that I buried fourteen distinct headaches in that garden in one summer at a decided gain in medical attendance. It is certainly fair, then, to put the labor in the garden as free, because it would have been spent on something in any event. Besides this, the crop from the garden was a real money return for my labor. Within the next thirty days we used on our table or gave away thirty-five heads of lettuce at an avarage price of five cents, or $1.75. Some of the plants were transplanted twice and the space occupied by the mature crop was about twelve by three feet.

I knew from former experience in the business that a garden could be made to pay. How much I resolved to find out, as soon as may be, and for this purpose opened an account with my little plot, which account was duly made up and balanced at the end of the year. Herein is set forth the *pros* and *cons* of the whole business.

I paid out for labor, seeds, tools, etc., just $6.61. There was sent to my table, between June thirteenth and November first, vegetables of all kinds to the value of $25.82. Besides these vegetables there were produced sixty-four strawberry plants (Jessie) worth at the time, as the Jessie is a new variety, at least two cents each, or $1.28. This made a total return from the ground of $27.10. Deducting the cash paid out there was just $20.49 left as the final result of my summer's work.

Pitiful little tale, not worth recording.
Think so?

It is small—only a trifling matter of $20.49. At the same time, $20.49 is $20.49 and most of us would accept it with a cheerful heart. No one was the worse for my partnership with nature. It was bran new money and came out of no man's pocket. Our table was supplied with vegetables for over four months, so that no purchases (except one quart of onions) were made at the stores for this time. Besides this, notwithstanding a rather poor season, the vegetables were of a far better quality than could be purchased anywhere. As an illustration of this I may confess without a blush that I ate nine cucumbers a day for several weeks in entire safety and complete satisfaction. To buy so many for one person would demand considerable moral courage, not alone for the price, but from the doubtful character of cucumbers two days old. Mine often reached the breakfast table in less time that it took to make the coffee—hence their beautiful innocence.

Did it pay? Would it not have been better to lay the lawn to grass, and to trundle a lawn-mower or toss the light tennis ball? Can't say. I am not a Tennis Courtier. But I do know that out of the ground comes health and wealth. Will you bring the children up forever on canned goods, when they might pull peas and good times out of the same ground. A home garden, even if it be only a patch like an extra large handkerchief, may in many a man's life-account make all the difference between profit and loss between a dish of greens and a lot in the cemetery.

From a recent report of the Bureau of Labor Statistics in Connecticut it appears that forty-six families representing twenty-nine trades and living in different parts of the State, were, at the time of the report, financially unhappy. The total income of the fortysix families amounted in one month to $2,475.36.

Their expenses for the same time reached $2,760.39—
an average loss for each family of about six dollars.
  Income, $2 a day ; expenses, $1.98—happiness.
  Income, $2 a day ; expenses, $2.01—misery.
  There is a fine flavor of the *Castanea vesca* about this
ancient joke, yet under its humor is a grim truth.
May not the truth about these Connecticut families
and many another in like unhappy plight be found—in
the garden? The report does not say that these forty-
six families had gardens, yet it must be observed that
Connecticut is a State of small towns, and that a very
large part of the population live in houses having
more or less ground. The report does state, on the
other hand, that one of the large items in the expenses
of these families was for vegetables. These people lost
$1.50 a week for each household. Could not that
amount have been taken out of their gardens?
  It will be said that even if the people had a bit of
ground it would not pay to cultivate it, that there was
no time for the work in the garden, and that the interest or rent of the land would be too high to admit
of profit. Is that so? No. It is just the other way.
In small towns where a house and ground are leased
the rent is for the house, and the land is practically
free, because if the value of the land becomes too high
the houses cover the entire ground or a large part of
it, as in a city block. Besides this every house must
have space for light and air, and this space is or ought
to be a garden. Certainly if a man has space round
his house, and he suffers it to go to waste when it
might produce food, he is morally responsible for the
loss he puts on other men by reason of his unpaid
debts. As for the time, it is only a question of getting up earlier or dropping the paper to tickle the
ground with a hoe—and better business any day.
Seeds, tools and fertilizers are cheap, and if there be a

will, there may be both time and a way. All of these families depended in part for their support on the labor of their women and children. Any twelve-year-old boy or girl could have put something from the garden on the family table and been all the better for it. As wages go it may be a question whether any young woman staying at home and minding the garden for three months would not earn more money than she could get in a mill or store. Her house would be her market, her family her customers, and she would reap all the profits.

There are the young folks. Let no young man or young woman fancy their education complete without some knowledge of the growth of plants. Gardening is an accomplishment worth far more than the ability to struggle through a sonata on the piano— more worthy of a lady too. Labor with the hands, in partnership with nature, on the sweet and honest earth, is worthy any gentleman. If there be any among you having a boy or girl, halting between their school books and a wish to climb the Golden Stair, let him consider whether it be better to have hands browned in the glorious sunshine, a face freckled by the blessed winds, clear eyes keen for out-of-door sights and pleasures, a little dirt, beads of salt perspiration, perhaps with a touch of the backache, a jolly appetite and a grand power of sleep, or white hands folded under a coffin lid.

It is the sum of these things that moves me to here set forth how any man, woman or child having a bit of ground may use it for their best health and the greater glory of their dinner table and pocketbook.

CHAPTER III.

TIMES AND SEASONS.

THE moment we begin to study the lives of plants we are brought face to face with the universe. We cannot consider the very small without considering the infinitely great. The life of every plant in our garden hangs on the laws governing the movements of the planets. Every melon vine that lifts its yellow cups in the warm air looks to a star for inspiration, and its fruit claims acquaintance with the sun by the blushes in its melting heart. The roll of our planet to the east divides the day into light and darkness, wherein plants grow and sleep, work and rest. The swinging of our old earth around the solar spaces divides the garden into a time for birth and a time for death. These things we must understand before we plant a single seed.

In all our country the year is divided into two parts, the growing season and the season of slower growth or of complete rest. On the twenty-first of December the sun casts its longest shadows at noon. After that day the days slowly grow longer and the nights shorter until the twenty-first of June. Then the shadows are shortened at noon, the days at their longest, the nights very short. Soon after there is, in the Northern States, a perceptible shortening of the day, which continues till the longest nights late in December again.

All plants, whether out-of-doors or in the house, are susceptible to these universal changes. The spring or growing season begins on the twenty-first of December. The winter or season of reduced growth, maturity, and sleep or death begins on the twenty-first of June. During the growing season the amount of light steadily increases and the plant thrives, because, as it grows, it demands more light. When it is going to rest or approaching maturity, or death, it requires less and less light. The lives of all plants are, therefore, dependent on the changing amount of light resulting from that motion of the earth that gives us the seasons.

At first sight it may seem that this cannot be true. In the latitude of New York nothing begins to grow out-of-doors before April, and all the garden plants are dead or asleep long before December begins. This, too, is true, yet I hope, in treating of certain of our common vegetables, to show you a number of experiments that will prove that this division of the year into two parts is correct. In California the growing season begins in November and ends in May. This, it is easily seen, is merely a variation, depending on local causes, of this same law of plant growth.

To get the best results from our garden vegetables, we sow the seed in the growing season and let the crops mature in the resting season, and this holds good, both under glass, where, as far as temperature is concerned, we are independent of the climate, and in Florida where there are no frosts. After the "return of the sun" we can sow any seeds, provided it is warm enough either naturally or in a greenhouse or sunny window. The days are growing longer, there is more and more light, and the plant finds its growing stature met by increasing light and (out-of-doors) increasing heat. Seeds of certain plants can indeed be sown in the fall or under glass as late as December, yet they are struggling against the solar tide and are never so thrifty as when the "young flood" of the year sets in. Peas, that are short-lived plants, may be sown out-of-doors in August, and will mature a crop before frost, yet they do not display the vigor shown by plants of the same class planted in April or May.

In the garden and out-of-doors the growing season is, of course, dependent also on the temperature. The spring, even in Vermont or Michigan, begins with the turn of the season, yet it is practically delayed out-of-doors for several months, or till the increasing light and sunshine raises the air to temperatures suitable for growing plants. In like manner the end of the growing season is forstalled by the return of cold weather many weeks before the actual end of the season at Christmas week.

The first thing we have to decide is this :—Can we take advantage of the actual beginning of the growing season without regard to the beginning of spring out-of-doors? If we can do so, we shall find a very great gain in the matter of early vegetables. Our object must be to get the greatest possible result from our garden, and to do this we must begin the spring work as soon as the season really opens. For instance, the tomato is a native of a tropical climate, where the warm weather begins early and ends late. In the latitude of New York the season out-of-doors is not long enough to bring its crop to maturity. We, therefore, gain time by starting seeds in the house soon after the season turns. The plant being sheltered from the cold and finding the spring really at hand, grows rapidly, and by the time the warm weather has arrived out-of-doors, is several inches high and well advanced in its natural life. In May we remove it from the house to the garden, and it thus reaches its full maturity even in our short out-of-door season. Cultivated in this way it produces its crop in August and September, whereas, if its seed were planted out-of-doors, it would be cut down by October frosts with only half a crop on its branches. It might be thought that the seed could be sown in the house in November and the crop thus be made to mature in June. This would not work, for until the spring really begins, it is nearly useless to attempt to sow seeds in even the warmest house. The young plants would be struggling against the solar tide and wasting their lives for nothing.

These facts in regard to the divisions of the growing year point to the first lesson in all horticultural work. Whatever we do, much or little, whether our garden be large or small, we must be forehanded. We must always look six months ahead, always lay out our work weeks in advance. If we wish tomatoes in August, we must plant the seed in March, and this means soil to put the seeds in, and to have good soil in March we must prepare it in November. Forethought and forework are essential to success in home gardening.

To show what is meant by planning the work in advance, I may from my journal give a few notes as to what was actually done to prepare for the season of 1888. By the first of November the last of the celery in the garden had been taken up, and the ground was left clear of all perishable crops. At odd moments the soil was spaded up and left rough, thus exposing it to the frost and rain to kill the eggs of insects and the seeds of weeds, and by December the out-of-door work was fairly over. The five strawberry plants had increased to over sixty, set out in rows about a foot apart each way. The garden was so exceedingly small that it was necessary to crowd them together to gain room, the intention being the next year to allow no runners to grow. On the twenty-first of November the ground was frozen hard and the strawberries were covered with dead leaves. Over this was laid some boughs and sticks to keep the wind from blowing the leaves away. A friend had given me in October a hundred currant cuttings, half Red Dutch and half White Dutch. These had been carefully set out in a bed by themselves, and were covered with leaves and brush about the twentieth of November. A few dozen grape cuttings, also a gift, were placed in a wooden box and buried two feet deep in the ground. Meantime I had sifted two barrels of good soil, mixing it with bone-meal, wood-ashes and guano, and stored it in the cellar. Then the snow came and the season was at an end.

CHAPTER IV.

PLANS FOR WORK.

IT is curious to note how quickly, after the season turns, there are visible signs of spring. No matter if the snow does fly in the north. Far down in the south the spring has landed on our coasts. The wave of green grass will cover the land, creeping up the Mississippi Valley, and stealing along the Atlantic Coast. The snow that covers the northern half of the country will retire slowly, sometimes pausing, sometimes advancing far down south, only to retreat farther than before. The buds soon begin to swell along the Gulf, and hints of spring are in the air. In the north we can only watch the slowly lengthening twilight over the snow-clad hills. By the tenth of January there is a perceptible lingering in the sunset colors, and, if we mark the spot where the sun goes down, we see he is already well started to the left, or south.

My home is in the neighborhood of New York City, and these notes from the journal of my handkerchief garden refer to the out-of-door seasons there. If you live south of Washington or Cincinnati, the work would come about ten days earlier. If farther south, still earlier. The best plan is to observe nature yourself. If the spring begins in your neighborhood in March, then you must be getting ready in January. The out-door season here begins in April, so my work begins in February.

First of all—books. There is no greater pleasure

in midwinter than to think over and plan for next summer. Nothing better for guide and companion on long winter evenings than a good book. Buy if you can, borrow if you must. Next to a good book, and, in some respects, better, is a good horticultural paper, and THE AMERICAN GARDEN is one of the best. If nothing better can be afforded, get the seedsmen's catalogues. Some of these can be obtained on application, others cost from ten to twenty-five cents, and are worth the money. Some of them are positively delightful. The descriptions of the seeds are so appetizing, and the pictures so inspiring, that we long for summer to come that we may enjoy these enticing heads of lettuce, these phenomenal beans, and gaze in awe upon our own monster squashes. The wise home-gardener will, of course, read the seedsman's catalogue with a dignified reserve in regard to some of the more bewildering pictures and their legends, yet he will read to learn, for nearly all these books are well worthy careful study, for the fund of valuable information they contain. At the same time, don't read less than three, each from a different city. Philadelphia, Boston, and New York, for instance, and read to compare opinions on the various standard sorts of vegetables.

Of books, I would recommend the following: The "Home Acre," a series of eight articles, by the late E. P. Roe, and published in Harper's Magazine, beginning in March, 1886. As a handkerchief garden should include some fruit, "Success with Small Fruits," by the same author, should be procured. It is a fine book, and beautifully illustrated ; price, $2.50. "Harris' Gardening for Young and Old" is a good book for general purposes ; price, $1.25. Bailey's new "Horticulturists' Rule Book" is full of valuable rules, recipes, methods, etc., for gardening folk; price, $1.00.

Henderson's books are all first-rate, "Gardening for Pleasure," for your purposes, being perhaps the best. "Gardening for Profit" is also very good, though designed more for market gardeners than for handkerchief gardening. If possible, I would also have Mrs. Treat's "Injurious Insects of the Farm and Garden." It's a handy book to have in the house, in case of war within your borders. These books, and all others on gardening and rural life, may be purchased of the Garden Publishing Company, New York. Reading, selecting varieties to plant, and planning out the work, may well fill the first six weeks of the year. By all means make a map or plan of your grounds, drawn to scale, so that you can see exactly what can be done. I did this in January, and laid out on paper every row of plants I wanted, before ordering the seeds. Afterwards the plan was of great value as a guide in using all my garden space to the utmost economy. It was also useful in economizing seeds, and in serving as a guide as to the quantities of each kind to be bought. It is also a good idea to preserve these plans to compare with the actual results when the crops are gathered. We are almost certain to plant too much of some things, and the plan will be a guide in next year's purchase of seeds.

There is also another advantage in making a plan of the future plantings. To get the greatest possible return out of the soil we must produce two crops each year, or three crops in two years. Suppose your garden is the usual city back yard, 25 feet wide and 60 feet long. Out of this bit of ground you must wring in one season all it is capable of producing. The ground must be stuffed with plants—not a foot, not an inch being wasted. If lettuce plants will mature when planted 12 inches apart, radish or some

PLANS FOR WORK.

short lived plant must grow between. If a cucumber vine covers 6 inches at one time and 6 feet at another time, spinach must occupy the space not used by the vine while it is small. If the early pea vines bear fruit in July, then white turnips must mature in the same ground in October. Like a circular race track, a garden to pay must consist of a series of "laps," one crop overlapping another and the soil bearing two crops between frost and frost.

To make this clear the ac-

I 6 x 12 ft. First planting bush beans 6 rows	⊠ Seed bed for celery. VI 11 x 11 3 rows early cabbage 3 rows lettuce 6 rows radish	XI 6 x 21 10 rows Early peas
II 6 x 9 Early beets 8 rows	VII 11 x 14 3 rows cauliflower 3 rows lettuce 6 rows spinach or radish	
III 6 x 7 Early carrot 5 rows		XII 6 x 7 6 rows onion sets
IV 6 x 12 Second planting bush beans	VIII 11 x 7 6 hills cucumbers or squash 6 rows radish	XIII 6 x 7 Early turnips
V 6 x 20 First planting spinach 20 rows	IX 11 x 7 Second planting cucumbers or squash, 6 rows spinach, radish, or 3 rows lettuce X 11 x 16 Early potatoes	XIV 6 x 25 10 rows second planting peas

companying diagram shows how the first lap is made. The entire space, 25x60, is laid out in three beds, with two footpaths, 12 inches wide, between them (and a good plan is to make these paths of plank). One bed, 11 feet wide in the centre, and two of 6 feet on each side, enables you to reach every plant with a hoe from the paths. On the plan these three beds are divided off into smaller beds, each for one or more crops. Beginning at the upper left hand (northwest) corner, the first bed is to contain six rows of the first planting of bush beans. (The area of each bed is marked in the beds.) Next, south of these, are eight rows of early beets; next come five rows of early carrots. These three beds are planted as early as the weather permits. The next bed, No. IV., is left empty till ten days after the first sowing of beans, and makes the second planting of beans. No. V. is planted thickly with spinach, in rows 1 foot apart. As soon as the plants are three inches high pull half of them out and send them to the cook. The little, half-grown plants make an excellent dish, and the plants left behind have more room.

I tried this plan last year, sowing spinach very thick and making the first thinning when the plants were very small. It took 500 plants to make a dishful, but they were delicious. Two thinnings and one final picking of the half mature gave a very large return in a very few weeks from a small space.

Bed No. VI. is to have four rows (north and south) of cabbages, the first row 1 foot from the path, the next two 3 feet apart and the last within 1 foot of the other path. Between each row is a single row of lettuce plants, and between lettuce and cabbages can be planted six rows of radish or spinach. Don't be afraid; you cannot have too much. Pull it up as

PLANS FOR WORK. 23

soon as it crowds the lettuce, and pull up the lettuce as soon as it crowds the cabbage. In both cases the crops will be ready for the table. Bed No. VII. is arranged in the same way, except that cauliflowers stand between the radish and lettuce. Bed VIII. is to contain six hills, first planting of cucumbers, and the whole space is filled up, except near the young cucumber plants, with spinach and radish. Bed IX. is the same idea applied to late cucumbers. If it is preferred, summer squash can be

	7 tomato plants	
I 6 x 12 ft 20 Hills late sweet Corn	**VI** 11 x 11 Cabbage followed by late Turnips	**XI** 6 x 21 40 Hills late sweet Corn
III 6 x 9 2d planting late corn 18 hills	**VII** 11 x 14 Cauliflower followed by Ruta Baga or Parsnips.	
IIII 6 x 7 Celery		**XII** 6 x 7 Late Cabbage
IV 6 x 12 Celery	**VIII** 11 x 7 Cucumbers followed by late Spinach	**XIII** 6 x 7 Late Cabbage
	IX Cucumbers followed by late Spinach	**XIV** 6 x 25 Late Potatoes
V 6 x 20 Late potatoes or Celery	**X** 11 x 16 Late Spinach	

used in the same way in place of the cucumbers. Bed No. X. is for potatoes that have been forced in the house. Bed No. XI. is for the first planting of peas, ten rows. No. XII. may be used for onion sets, six rows, and No. XIII. will carry early turnips, six rows. Bed No. XIV. is to be kept about ten days, for the second planting of peas. The little square at the north end of the plot is a seed bed for celery.

In sketching out in winter such a plan of work for the summer, you must look beyond the early spring and arrange for the crops that are to follow the early spring plants. With this is a plan of the same garden, showing what should follow in the various beds.

For instance, bed No. I. may be followed by late sweet corn as soon as the peas are gathered. Bed No. II., in like manner, may be used for the second planting of sweet corn, the beets to be consumed as soon as half grown. The little bed of celery plants is to be cleared out before the tomatoes begin to crowd them, and the young plants moved to beds III. and IV. The tomato plants are shown in a row at the upper end of the lot. There will be ample room to get them in there, and if not, you can well afford to sacrifice a plant or two of the first crops in beds Nos. I., VI. and XI. All the beds are plainly marked with the second crops, and you will find it well worth while to compare the two plans, as they show how crops may be made to "lap" and how to get the greatest possible return from the ground.

My own garden was of a somewhat different shape, yet I made careful sketches of the proposed crops, and in the summer of 1888 actually carried out, with some varieties, the succession of crops shown on these two diagrams.

CHAPTER V.

STARTING THE GARDEN IN THE HOUSE.

TO get the greatest possiblé return out of a handkerchief garden, we must forestall the growing season. It is the early potato that costs. We must gather our crops when they are high-priced in the stores, and thus credit the garden on a "bull market." It is better to stop buying lettuce when it is seven cents a head than have to wait till it is down to three cents. The market gardener's chief profit is always in these forwarded and high-priced crops, and we must be equally sharp after every early penny that grows in the garden. It is often thought that only those who have green-houses and hot-beds can thus hasten their early crop. Glass is always a great help, and it pays to use it, yet for a small home garden it is not necessary. Every house has one or more sunny windows, and these make the advanced garden where the early crops may be started.

Can't have troublesome plants making a slop and dirt in your parlor?

Not the slightest need of it—if you know how Besides, a neat box filled with young cauliflower plants is rather pretty and suggests the spring, long before the snow has gone. Even a box of young potato plants, thrusting up green fingers towards the light, may be quite a picture. Visitors will be sure to look at the cheery bit of greenery and ask in all innocence the name of the odd looking plants.

Spoil your carpet and fade your curtains?

Spoil your children too? What's the use of carpets, if the grand sunshine full of health and cheerfulness is to be shut out? Better burn your old carpets and let the sunlight fall on bare floors. Better a row of plants in your parlor window, and delicious summer cabbages in July, than a best room shut up dark while the pale-faced children mope in your stuffy north kitchen. These things are not for fun. Its simply good business to hasten the handkerchief harvest and thus reap the big profits. Besides this, you have been carrying the children for weeks on canned goods. A taste of the first salad from the garden will save the doctor's bill and tone up every little stomach in the most encouraging way.

Our house faced south-east, and this gave us four sunny windows down-stairs, two facing south-east and two south-west, in rooms warmed by a furnace. There was also one sunny window up-stairs, in a room partly warmed by a chimney, and the spare heat from the hall. In the two kitchen windows shelves were put up, and in the parlor and dining-room small narrow tables covered with cretonne were used. A good idea in putting up shelves for plants in the lower part of a window is to have a piece of shelving made to fit the window and about eight inches wide. On the edge of the window seat screw two small brass hooks. Opposite these, on the upper side of the shelf, fix two screw-eyes. On the under side of the shelf in the middle fasten a single iron bracket. To fix such a shelf in place, put the screw-eyes over the hooks and the bracket prevents it from falling. Such a device saves all nailing into the wood-work and the shelf can be unhooked and removed at any time in a moment, in case the maid wants to wash the windows. I first saw this little notion carried out by my friend and neighbor, Bronson Howard, who is as clever with tools as with the pen. Two shelves are enough in each window, and I found a lath, covered with cretonne, as a concession to the æsthetic, and nailed across the window frame on a level with the top of the lower sash, made a good support for light window boxes.

Flower pots will be needed for the windows, and it is well to have a few of different sizes for some of your work. It is, however, very much cheaper to use wooden boxes. In my own experience I found there is nothing better than a bundle of laths. It cost delivered only 30 cents, and out of it I made dozens of plant and seed boxes of all shapes and sizes.

The accompanying sketch shows one of my window boxes. It was made by cutting ten laths into lengths

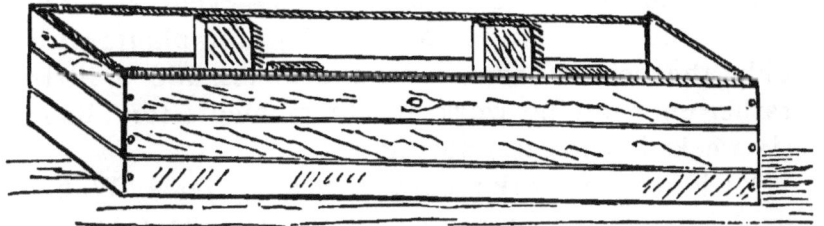

of 2 feet 10 inches, and nailing them together in two sets of three each (placed side by side) and one of four laths. They are fastened by the short crossbars and then the three sets are nailed together and the ends filled up and all made secure with small wire nails. To prevent splitting I keep the laths in a damp place till ready to be used. Such boxes are of a convenient length for the window and will just fit into a cold frame measuring 3x6 feet. For small seedlings I made boxes four laths wide and two laths high, and as long as the window sash, so that they would rest on the bar in the middle of the window and the top of the sash. If you wished, the side next the room could be covered with cretonne or painted some dark color, and then filled with the pale green of young lettuce plants, they would look well in the best room of any home.

From my experience with such window boxes, six boxes, occupying three sunny windows, would be sufficient to supply all the early cauliflower, cabbage, lettuce and tomato plants needed in a home garden intended to supply a family of three adults and two children. Supplemented with cold frames, covered with either glass or protective cloth, they would easily carry one thousand plants, or more than enough for a dozen handkerchief gardens. My experience is that, even without frames, all the plants you need for your home garden can be raised in your windows without a single cent's extra cost in the way of fuel. Your home must be warmed in any event, and the same heat will bring on a crop of young plants with only the cost of the seeds, the boxes, and a little rather entertaining work at odd moments for about six weeks in the early spring.

My journal of work records that Early Jersey Wakefield cabbage and Extra Early Erfurt cauliflower seeds were planted in boxes on February 22d, and the Early Snowball cauliflower planted March 9th. The first lot of plants were transplanted into other boxes by the middle of March, and were removed to the cold frame early in April, and were set in the open ground April 27th. The second plantings came a little later, and the first cauliflowers were placed on the table on July 4th, while the last were eaten on July 22d. The first cabbages were cut on July 15th, and they lasted well into August. To those who have never tried it, early summer cabbage, just beginning to head and fresh from the garden, will prove a new dish. You may have eaten something so-called and thought it very good. You haven't really been there, unless you have a garden of your own.

STARTING THE GARDEN IN THE HOUSE. 29

On February 28th I cut up a few choice potatoes and placed the pieces in soil in a window-box by a south-east window in a room up-stairs where there was no fire. On very cold nights the box was placed on the floor near the chimney and covered with a newspaper. On April 16th the plants, now six inches high, were transplanted to a sunny corner in the garden, and the first potatoes were sent to the table on June 27th.

Lettuce was planted on March 7th, and set out in the garden on April 3rd, and the first heads eaten June 11th. Tomatoes with me were a failure, owing to damping off at the time of the blizzard in March. Still, plants from seed planted March 8th made good plants, and would have been set out in the garden in May, had it not been for an accidental upsetting of the box that compelled me to buy plants of the nearest florist. I did enough, however, to prove that tomato plants can be raised in the house without the slightest trouble. Among other things, I found that peas can be forced in the house by sowing in boxes the last week in February, and transplanting to the ground when about four inches high. The crop from these transplanted peas came in about three days before peas planted in the ground, as early as the weather permitted. The gain was slight, yet in a favorable season I think it would be even better and would pay to do, if you want extra early peas. The plants were set out quickly and with no particular pains, and not one died.

Not having any material for a hot-bed or even glass for a cold-frame, I made a frame on the south side of the house protected from the north-winds, and for sash used frames made of two old screen doors, covered with a heavy grade of protective cloth. Under this frame I forwarded potatoes, cabbage, cauliflower, tomato and lettuce plants, that were afterwards transferred to the garden. The frame measured about 2 feet 8 inches wide and about 18 feet long, and held many hundred plants, in fact far more than the garden would contain, and two-thirds of them were given away or sold to the neighbors. On very cold nights the frame was covered with old gunny bags as an extra protection, and on all sunny days the frame was left uncovered, except in high, dry winds. While not as good as glass for some purposes, this protective cloth answers very well for forwarding early plants. Cabbage, cauliflower, potatoes, peas and lettuce under it did very well. Tomatoes not quite as well, and another year I would keep tomato plants in the window or under glass sash. The frame cost for lumber 50 cents and for the cloth $2.16 = $2.66.

Another advantage of such a frame is the protection it affords to young squash, melon or cucumber vines. Seeds of squash planted in the frame came up among the other plants, and as soon as they needed more room, the nearest plants were removed, and finally the frame was wholly taken away and the vines spread naturally over the ground.

These things were easily carried on at odd moments through February and March, and in April the regular out-of-door work began. By May 7th the frame had been taken away, and its contents had been transferred to the garden. None of the work in the house or about the frames took more than an hour or two at any one time, and usually the time spent over the work did not exceed fifteen minutes twice each day, say once early in the morning and once towards night. Often it was very much less. None of the work required much strength or skill—only a

STARTING THE GARDEN IN THE HOUSE.

little patience and the right hand's turn of work at the right moment.

Among other notions for forwarding early plants is a plant-hood or folding-tent to put over small plants. The device received the name of *The American Garden* Cosey, and as it may also prove useful to others, a few directions for making one or more may be given.

The materials of the cosey are protective cloth, common laths or other light wooden sticks, and common carpet tacks and any stout twine or small cord. The first one made by the inventor was made out of four laths and 67 inches of a heavy grade of the cloth. Spread open on the ground to cover plants it protected a space 4 feet long and 14 inches wide, giving ample room for a mature lettuce plant or strawberry plant in bearing, or any young plant not over 18 inches high. When shut up it could be put in a space 4 feet long, 18 inches high and $1\frac{1}{2}$ inches wide. A dozen would be a light load for one hand.

To make a single cosey for protecting a few plants, cut a piece 19 inches long and from this cut two triangular pieces, each 17 inches wide at the bottom. There will also be material for one more in case another cosey is made. 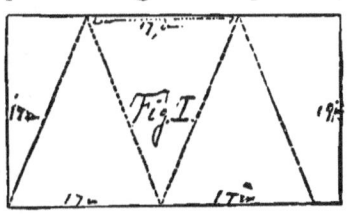 The dotted lines in Figure 1 show how the cloth is cut, the fabric being 36 inches wide.

These two pieces will make the end pieces. To make the cover, cut a piece of the cloth $1\frac{1}{3}$ yards long. For the frame, use four good straight laths. Place them in pairs and join each set with a crossbar at the end, 18 inches long. Nail firmly at the corners and put in a short brace at two corners to keep the frame in shape. Figure 2 shows one of these frames.

Place the cloth on a table and then lay the two frames, with the crossbars together, one over the other on the cloth near one edge of the cloth. Then fold the cloth over the two frames and tack the edge to the lower edge, leaving about half an inch of the wood exposed. This is to prevent the cloth from touching the ground. Leave the ends (for a couple of inches) free till after the end pieces are put in. This done, turn the two frames over, stretch the cloth tight and nail it to lower edge of the other frame as before.

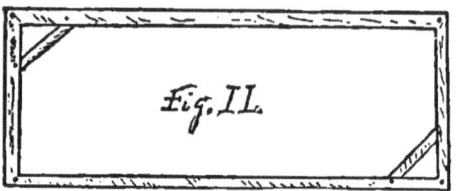

Then tack the cloth to the upper bar of each frame, placing the tacks on the wide part of the bar. Lastly, put in the two end pieces, lapping them over the frames and leaving a loose flap 2 inches wide at the bottom. When the ends are tacked on, finish the cover by drawing the ends over the edges of the end pieces to make a neat join and tack the ends down.

When finished the cosey can be opened and will stand alone, making a rain-tight hood 4 feet long and 14 inches wide. It can then be placed over plants, gently pressed into the soil to fit tight round the sides, and a little soil can be thrown on the flaps at the end to exclude the air. This is a single cosey for a few plants. To cover more, say a space 8 feet long, make two coseys and close up only one end of each and then place them end to end, the two open ends meeting and thus making a continuous hood of the two coseys. To cover the crack between the two coseys, let the cover of one extend 2 inches beyond the open end. Set the first cosey in place over the plants first, then place the other in position, letting the flap cover the crack all round. In this manner a long line of coseys can be used, each one having a flap to fit over the next with only the two end ones having the end pieces. Where the ends are left open in this way a stout string must be secured across the ends to keep the frames from spreading, in fact, acting as a tie-rod to hold up the roof. Figure 3 shows two coseys placed together.

The cosey is one of those little notions that often prove of great value in many ways. It can be used to forward early crops in the spring and to protect late crops in the fall. It is wide enough to cover two

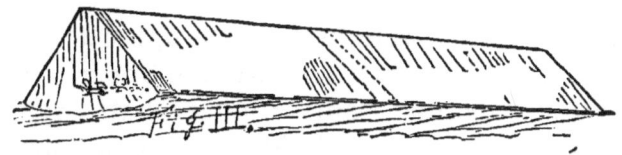

rows of early beets, carrots or radish, or to protect a row of strawberry plants from the birds or to keep insects away from young melon vines. Such protecting hoods could, of course, be made of other shapes and sizes, but this size uses the cloth without waste, and the hood is easily picked up, carried away and packed snugly in the barn when not in use. The cloth will shed any rainstorm and will not mildew or decay in wet weather. It is also a protection against frost and is better than glass and only one-tenth its cost. Such a cosey, or even a long line of them, can be easily ventilated in bright sunshine by putting a block of wood or a stone under the edge, or, where two or more are used, by pulling them apart and leaving a small space between them through which the hot air can escape.

The "protective cloth," above mentioned, is made especially for such purposes, and is sold by seedsmen.

CHAPTER VI.

CLOSE CROPPING.

COMMON objection raised to a home garden is the expense and labor in planting and caring for it. My little patch, which measured a trifle over one-thirtieth of an acre, was a tough, hard clay, and yet, in addition to one man's labor for one day, it required only nineteen half days' work in April and ten half days in May. This included the preparation of the ground and planting all the early crops. In no instance was a whole day's work given to the place, and the time spent was usually from four to six in the afternoon. I think the entire work could easily have been done in ten working days. The entire outlay, aside from my own labor, including new tools, seeds, lumber, cloth and fertilizer was $14.79. The heaviest expense was for 200 pounds Mapes' fertilizer, $5.05 (delivered), and this was enough for the entire season, no other manure of any kind being used. Fourteen dollars and seventy-nine cents would buy quite an assortment of vegetables, canned and otherwise, at the stores. Did it pay? Was it worth the expense, labor and trouble? It certainly did pay, as I propose to show at the summing up of the season's work. Meanwhile it may be well to see what was obtained for the money. By the first of June the following vegetables had been planted at different times: four kinds of peas, four kinds of radish, upland cress, chicory, leeks, potatoes,

two kinds of spinach, two of onions, two of squash, carrots, parsnips, two kinds of beets, two of bush beans, four of lettuce, four of cabbage, two of cauliflower, one variety each of tomato and turnips, and three kinds of celery. Quite a bill of fare for a small place. There was, beside this, a small strawberry bed, a cutting bed of currants, and one of grapes, both of which received a part of the labor and fertilizer. In addition to all this there was a good flower garden, that demanded more or less attention, and produced a very large crop of flowers from the first of June till frost in October. The object of having this great variety of vegetables was twofold. In the first place, it is important to find out something concerning the character of the soil in a garden, and the only way to do it is to try many kinds. For instance, I learned that turnips and radishes were unsuccessful, lettuce and celery very successful, showing that the soil was best for the last, and that in the future it would be better to have more of one and less of the other. In the second place, to get the best return from a garden, attention must be paid to the daily bill of fare in the house. The selection of seeds and the planting must be so arranged that there is always a succession of things for the table, and not too much at any one time. Even with this great variety, we had in July eighteen heads of cauliflower ripe at one time, far more than could be used, and a dozen heads were given away to the neighbors. It was the same with summer cabbage, nearly two dozen ready at one time, a bigger crop than the home market could absorb. Other things came to perfection in about the right quantities, and the table was usually supplied with three kinds of vegetables every day through the summer. After the first of June no vegetables, except potatoes, were bought, and after the thirtieth of

June nothing whatever was purchased in the way of vegetables for many weeks, and had late potatoes been planted, the garden would have carried the house till December.

The labor spent through the summer was very light. Spreading the fertilizer, transplanting and raking over the soil to keep down the weeds, made the whole of the work, and none of it took over two hours on any one day. The crops were gathered every day, just before or after breakfast, and took only a few moments, when a little turn in the fresh air was more a pleasure than a task. The system of overlapping crops already described worked perfectly. On a plot 6x10 feet I planted Savoy-leaved spinach, and when well up set out very close together between the rows three dozen early Jersey cabbages. Six pecks of spinach were taken off the plot, the two first pickings being "thinnings." The spinach completely covered the ground and yet it was all cleared off before it interfered with the young cabbage plants. The final picking was like a transformation scene, the dark green of the spinach bed being in a few moments changed to the pale green of a cabbage patch. On another part of my grounds I planted spinach in long rows, and as soon as the plants appeared set rows of cauliflowers between the rows. The spinach was gathered as soon as the plants began to touch the cauliflowers. In another place a row of early cabbages was set out and on the same day spinach seed was planted between each plant. The spinach came up and was gathered before it troubled the cabbages. By using the spinach when about half grown I had an excellent crop of early greens on the same ground occupied by other plants. Afterwards, as the cabbages were removed, late sweet corn was planted, so that the ground actually produced three crops in one season.

To still further carry out the plan of close cropping I planted summer squash between the rows of peas (second planting) and found no trouble from interference, the peas being pulled up before the squashes wanted the room. I had also under way a trial of peas and beans (Laxton's Early and Early Mohawk), the beans being planted in hills between every other row of peas and the peas trained away from the beans. The experiment worked well. The beans were planted just as the peas began to flower.

Another experiment in crowding the land was to plant onion sets in rows and quite thick in the row, and to plant cauliflower plants between the rows when the onions were about six inches high. The demands of the table gradually used up the onions by pulling up every other plant, and the open foliage of the onion did not seem to annoy the cauliflowers. Finally the onions all disappeared, and then the cauliflowers ripened and were pulled up and made room for celery plants in August.

Lettuce plants set out from seed box in window were planted in April between young strawberry plants and were well headed before the strawberry plants began to run. Among other stray bits of information I picked up the fact that if bush beans are injured, as mine were, by early cold rains, it is perfectly easy to transplant the young plants to repair broken rows. In a home lot space is too valuable to allow broken rows to mature. It's better to transplant and use the space with something else.

Another bit of crowding was arranged in this way: Two rows of spinach were sown (east and west) near the fence and parallel with it. As soon as well up, early cabbages were set out between the rows. The spinach was pulled and eaten before it troubled the cabbage plants, and then tomato plants were set next the fence and trained up against it, and before the tomatoes were ripe the cabbage had headed and, when pulled, left all the space for the tomato plants. None of these crops seemed in the least degree incommoded by the others, nor did they fail to give excellent returns three times during the season. In ordinary garden planting the plants would have been set in different places at a loss of space and labor in caring for them. The ground occupied by this experiment measured just 19x3 feet, and it produced two pecks of spinach, twenty heads of early cabbage, and carried six large tomato plants that produced a fair crop of tomatoes. It is only by this system of close cropping that a handkerchief garden can be made to pay large profits, and it is the only plan I would ever employ in my own garden. If you mean to garden at all, do it in this way and wring from the ground all the return it can possibly give. Make the garden tell.

CHAPTER VII.
A DISH OF SALADS.

YOU have a little space at the back of the house. It is very small, so small indeed that it seems hardly worth while to use it. There's nothing in it, save a few rank weeds. The sun only shines there a part of each day. If weeds will grow, something better will grow. The actual surface may be only a bed along the fence, say 25x4 feet. Small as the border is, it can be made to keep your table in salads four months out of every twelve.

First of the soil. Dig it up and see what it is like. As nearly all our cities grow outward into the country, it often happens that the yards about the houses contain very fair soil. If it is thin and sandy, good garden loam should be procured. Two one-horse loads should be enough. The florists can usually provide it for about $2 a load. If it is very heavy and is wet for sometime after a rain and cracks when drying in the sun, it has too much clay, and this defect can be easily cured by the addition of about a barrel of sand from the mason's yard. If weeds grow, the soil is pretty nearly right, and can be made just right by the addition of manure from barn or stable. If this cannot be procured, use one of the standard fertilizers, together with ground bone or bone-meal and wood ashes. These things can be procured at the seed stores by the pound. If possible have the soil spaded up roughly and left in that condition all winter. In the spring the fertilizers can be spread over the surface and forked in as soon as the ground is dry. Four pounds of commercial fertilizer, one pound of bone-meal and half a bushel of wood ashes will make a good mixture. There should be about five pounds more of the fertilizer on hand for use at intervals during the growing season.

The thing to grow is lettuce. It is the most useful crop we can have, one of the most simple and easy of culture, and it is always acceptable on the table. If you can raise but one crop, let it always be some variety of lettuce. If you have a sunny window, you will be 'surprised to find how many dozen heads of lettuce can be gathered from this mere ribbon of ground. The cost, including the fertilizer, will be very small, and the only labor of any consequence will be in spading up the ground. This ought not to exceed one hour's labor about six times each season and the few moments' attention once or twice a day from the first of March to the last of October. Much of the time, however, there will be nothing to do for weeks beyond the gathering of the daily crop. The entire time spent in caring for the lettuce will probably be less than the time required to go every day to the store to buy your lettuce.

The first planting in a flower pot or small wooden box should be about the first of March. These plants should be set out in the little bed by the tenth of April, setting the young plants in three rows one foot apart each way. One-quarter of a ten-cent package of seed will be ample for this first planting. A six-inch flower pot will easily hold it, and if three dozen plants are set out in the border, it will be enough for the first crop. One of the best varieties to use is the "Boston Market" (or Tennis Ball).

About the twentieth of March make a second sowing of the same quantity of seed in a box or pot in the window. The young plants will be ready to set out in the garden in about thirty days. As the first crop is still in the ground, set these new plants in a small bed by themselves about three inches apart. They will stand in this bed till there is room made for them by the maturing of the first crop. As fast as a head is ripe, pull it up and send it to the table, and stir up the soil and set a new plant from the small bed. Leave none of the space idle and keep transplanting at every opportunity that offers. It will be found that the crop will mature faster at times than it can be eaten. In this case the heads can stand for a day or two without injury. Small as the bed is, it will carry in various stages from four to seven dozen or even more through the early summer, and will easily give one head a day for the larger part of the season. With care it will be quite possible to have a head a day from June first to October first, or even later. If there is any gap in the supply, it will come in July or August, when the warm weather causes the plants to run up to seed.

The third planting should be in the open border about the middle of April, transplanting three inches apart as soon as the plants crowd each other, and a foot apart when the young plants again touch each other. For the fourth planting, which will be out-of-doors, use the "Hanson" lettuce, and this should be sown by the tenth of May. Plant the "White Russian" on the first of June and twentieth of June. To extend the season, plant the "Tennis Ball" variety again on the first of August, and for the last time about the fifteenth of August. These last two plantings will carry the crop well into October and keep up the supply till the frost cuts the plants down.

The work of attending this series of crops is very simple. On the day following every rain, break the surface of the soil round the plants by raking it lightly. A small steel rake, such as is often sold in "children's sets" of tools, will be found useful. This raking will also keep the weeds down, but if weeds do appear, rake them up as soon as possible. In this way all the hard work of hoeing will be saved. About every two weeks through the season sow some commercial fertilizer thinly over the surface of the ground just before a rain, and rake the soil gently to cover it. If very dry weather comes, shower the plants thoroughly on a bright sunny morning about twice a week.

It will be seen that this is very high as well as small culture, and it will be found to be very profitable. Such a little border should carry four crops of at least five dozen heads each, and even at three cents a head, should save $3.60 on your grocer's bill. Trifling little return you think? It is small, but the bed is very small. It may cost a few moments' trouble, and a little something for seed, soil, fertilizer, pots, etc. It may even cost more than you get the first year, but another season you should do much better. Try it and you will be convinced that it will pay, because the lettuce, with care, will be superior to any heads you can buy at any price in the stores. Fresh lettuce is one thing, store lettuce quite another.

If a little more space can be used, say one or two square yards more, sow in the early spring seeds of parsley. It will be very welcome to the house-mother all through the latter half of the summer. To extend the season take up some of the best plants in September, and they will grow in pots or boxes in a sunny window well into December, and furnish flavoring for soups, or dressing for fish. Another good plan is to buy a package of any good celery seed and to sow it thickly in rows about a foot apart. As the young plants come up pull the larger ones as you may wish them for the soup-pot. Even when the plants are only a few inches high they make excellent flavoring for soups. Another very useful plant is the new Upland Cress. Sow half a paper broad-cast in a little bed, and as fast as you want it for dressing a dish of fish, pull up the larger plants. A little later, cut off the larger leaves as wanted. A few of the plants set out in the border in a row, and about eighteen inches apart will extend the supply. The first sowing should be in April and a second sowing late in May.

If the family is small and less lettuce is needed, it will be a good plan to omit one row of the lettuce next the fence, and to set out in April a pint of onion sets. Plant them quite thickly, and by the first of June they can be pulled as fast as wanted for soups and stews. Pull every other one along the row, and then every other one again. In this way, in the course of a month, they will be slowly consumed, and those that remain longest in the ground will have room to grow. In crediting your garden with these small "stew-greens," parsley, cress, onions and celery, find out the price at the stores. They are usually sold in mixed bunches, several kinds in a bunch, at from two to five cents a bunch. These may seem trifles, yet they will save many a trip to the store, and many an odd penny that goes to make up a dollar, and help wonderfully in piecing out a "picked-up dinner."

Another useful plant is the Fetticus or Corn Salad. It can be sown early in the spring, and is ready for the table in about six weeks. Another plan is to sow it in rows a foot apart in September. When the ground freezes it must be covered with leaves or straw, and on approach of severe cold weather it must be covered six inches deep. It is uncovered early in the spring, and is ready to cut in a very few weeks.

CHAPTER VIII.

WHAT TO DO WITH A CITY YARD.

GIVEN a city lot and we have an area of 25x100. The house occupies usually about 40 of the 100 feet, leaving an open space in the rear of 25x60. Here the weekly wash must dry, and for this purpose there must be grass. The maid, when in the garden hanging out the clothes, would be heavy of foot on lettuce or roses, and so it usually happens the back yard produces nothing but grass. The usual plan is to stretch the lines across the yard from the fence, and, if the wash is large, the whole of the line is occupied close up to the fences on each side. This makes even a narrow border round the edge of the yard almost useless, and neither flowers nor vegetables are ever attempted. A better way would be to measure off the first six feet of the yard, next the house, for the whole width, and lay it with brick or stone for a walk or out-door sitting-room for summer evenings. Then lay off a space for grass in the center 17x48 feet. This would leave a border four feet wide on each side, and a border six feet wide at the opposite end from the house. The space in the center would be for the use of the maid on Monday, and for a pleasant play-ground for the children on other days. It would be also a lawn and a walk, from which to tend the borders. Instead of carrying the clothes line to the fence, have a post at each corner of the grass plot.

Too much trouble for a few heads of lettuce. Think so? Try it and you'll be glad you did try it.

Such an arrangement of a city yard would give three borders, one 6x25, and two, each 4x48 = 534 square feet. If the wash "is sent out," more space could be gained by making the two side borders each two feet wider. It would not be well to make them wider than this, as six feet is about as far as you can conveniently reach with a hoe or rake while standing on the grass. Many city yards that I have seen in New York are arranged in this way, except that there is a stone-covered walk eighteen inches wide around the grass plot, and leaving a very narrow border, often only a foot and a half wide, next the fence or three sides. Such a walk is a waste of room, for the grass plot can be used for a walk at a wonderful gain in comfort. No man has yet invented a carpet equal to grass for feet weary of city side-walks.

City yards are often used for flowers or for a few vegetables, and sometimes with ill success. There are two reasons for this. One is that the soil is usually poor and thin or stiff with clay. You must have good soil, and this is neither very difficult nor expensive to obtain, as is explained elsewhere between these covers. The other reason is the want of sunlight. The tall houses on every side cut off the direct sunlight for a portion of the day. This is not a fatal objection, if the right kind of plants are selected. There are plants that will flourish in partial shade, and by using these very nearly as good results can be obtained as in the best country garden.

The first thing to consider is the aspect. Which is the sunny end of the place, which the shady part?

If the house is at the south end of the yard, its shadow will be on the walk and the warm sunny corners will be at the opposite end of the yard. The plants needing the most light will therefore go to the north end. As the morning sun is better than the afternoon sun, the west border (facing east) is the next most valuable place. The aspect is, therefore, of value in this order : First, the north border ; then the west border, and, lastly, the east border. The north border is best for tomatoes, cucumbers or melons, the west for beets or carrots, and the east border for celery or lettuce.

In such a very limited garden it will not be worth while to attempt a great variety of plants. It is too small to carry all that would be needed in a family of five, and the best that can be done is to have a few kinds only, and of these only those that will thrive in partial shade. One of the first things to set out in spring should be rhubarb. Six good roots planted near the north end of the west border will be enough. This is in the nature of a permanent plantation, and once set will last for years. The roots must be bought quite early in the spring when the rosy tips of the leaves are just showing above ground. The roots are usually cut into small pieces, and they should stand in the center of the border three feet apart. The soil should be made as rich as possible before planting, good barn manure being best. If it cannot be obtained, use commercial fertilizers at intervals during the summer, as the plants grow. Set the roots in the ground with the growing point just under the surface. As the plants grow let them spread as they will. Do not cut any of the stalks the first season. When the flower stalks appear cut them off, as they only tend to weaken the plants. Rake the soil round the plants after every rain, or as often as weeds appear. The first crop of stalks can be taken off the second spring. Pull them off with a sideways twist to break the stalk close to the root. If convenient cover the roots in the fall with coarse stable manure and rake it off clean in the spring, as soon as the frost leaves the ground.

If your space is crowded, a single row of lettuce might be put in front of the rhubarb plants in the spring, before the leaves begin to spread.

For the north end of the plot tomatoes will be useful. Buy the plants already started in pots. Six or seven plants can be placed at equal distances across the end of the bed next the fence. As they grow, it will be found a good plan to give them a large trellis or guard for support. One good way is to support a barrel hoop on three small stakes and to put the hoop over the plants so that the heavy branches will spread over and lean upon it, and carry the fruit above the ground. Another good idea is to get two light wooden strips and place one on each side of the six plants and support them at each end with stakes driven cross-wise into the ground. Several of the seedsmen advertise a very good tomato trellis hinged at the top, and ready for immediate use in the garden. The main thing is to keep the heavy branches off the ground, and a few sticks and a little gumption will do it.

Tomatoes are very cheap and it might be said that there are more profitable plants for a city lot. This is true, and yet it will be found an advantage to cultivate tomatoes, as the fruit is best when quite fresh. In the early spring, while plants are small, the first three feet of this border, next the grass, can be used for lettuce, spinach or radish.

If preferred this warm border can be used for cucumbers or melons (not both.) The cucumber is a vine that can be easily trained on a trellis, and in a garden where space is so valuable, it will be found a good plan to set up a trellis of galvanized wire fence-netting. It is about a yard wide, and only enough is needed to reach across the lot. It should be supported on blocks from the fence to leave a few inches clear space behind it. When the posts are on this side of the fence it could be nailed to the posts. Six hills of cucumbers planted close to the netting would fill the space, and the young vines, when they are once led up to the trellis, will quickly run all over it, bearing their fruit and flowers in the air, instead of on the ground in the usual way. The fruit will hang from the vine and ripen on the fence just as well as when lying on the ground. If there are more vines than will fill the trellis, let them spread over the ground in front of it.

The culture of the cucumber and its cousin, the melon, is very simple. Have the soil made rich and soft, and sow about twenty seeds in an open ring or a circular patch and cover thinly with soil pressed down firmly. As they appear, pull the weakest ones out. Wait a week or ten days and then pull up all except six in each hill or group. In this way the excess of plants serves as insurance against insects. Some will be sure to be destroyed, and by having too many the crop can be saved. The after-culture consists in keeping the ground raked after rains till the plants become so thick that nothing more can be done. The vines should be examined every morning and all the ripe fruit removed, as a single cucumber, left to mature and ripen its seed will injure the vine far more than two dozen cut when half grown. The White Spine Cucumber will be found a good standard kind.

It seems to be a law in plant growth that, if any

plant is allowed to mature its fruit and perfect its seeds, it is content and will make no special exertion to bear more fruit that season. If its flowers or half-ripe fruits are removed it endeavors to produce more. If those in turn are taken away it will again flower, and seek to produce fruit and seeds. This is very marked in the case of annuals, like the cucumber and sweet pea. If the flowers are constantly cut, the vine will bear a great many flowers and keep in bloom for several weeks. If the first flowers mature, and pods and seeds are allowed to ripen, the crop of pea-blooms will be very small and the time of blooming short. The more cucumbers you cut, the more you will have. Better cut your cucumbers every day and give them away, for the more you give the more you will have to keep. Selfishness never pays as a regular crop.

For the east or most shady border the best things to grow are lettuce and celery. Two crops of lettuce (see Chapter VII.) can be taken off the border before the celery is put in. Buy the dwarf kinds of celery plants of your seedsman, and set out the plants in a single row, about ten inches apart, placing the row in the middle of the border. The culture is very easy when the one idea on which it is based is understood. The celery is a plant that is greatly improved by growing in the dark. The tough, green stems become crisp and brittle in the shade, and any method by which the stalks are protected from the light will give good celery. A bunch of plants growing thickly together in a mass, will so shade each other that those in the center will be blanched naturally. The most simple way to secure the blanched stems is to cover them with earth. This is called "earthing" or "bunching up," and it is nothing more then piling the soil against the plants as they grow.

The young plants should be set out between the sixteenth of July and the first of August, and for the first month the culture consists in keeping the ground loose and free from weeds. If the soil is very dry and there is little rain, copious waterings twice a week will be found useful, as the celery is by nature a swamp-haunting plant and a great lover of water. In ordinary seasons and in a clay or peaty soil watering does not seem to be necessary. I have raised on a clay soil fair crops without it, though I cannot say how much better the result would have been if a hose had been brought to help the hoe.

For about a month after setting the plants they appear to stand still and to make no growth. They are really extending their roots, and as soon as cool weather comes in early October they grow rapidly. As soon as this growth begins the first earthing-up must be done. One way is to tie all the stems of a plant together in a bunch, by tying a string just under the leaves. Another plan is to simply bunch the stems together with the hand while the soil on each side is pulled up against the plants, to bury them about half their length. Two weeks later more soil is pressed up against the stems till only the tops are visible, banking it up into place with the back of the spade. The plan of tying together with a string is best, as it can be done by a boy very quickly and once tying saves all further handling of the plants, and causes the center stems to blanch even before covered with earth.

Another plan is to tie all the plants in a row and then to set boards on edge close to the plants, one on each side, and thus to exclude the light without earthing up. The boards are easily kept in place by stakes driven in the ground, and the boards tend to make the plants taller as they stretch up to find the light.

The boards should lean against the plants, and may be kept in place by simply piling the earth against them. White Plume and Boston Market are good white kinds, and New Rose a good crimson variety.

Another crop useful in such a small garden would be spinach. Two sowings in the spring and one in the fall would be best, as city yards are apt to be intensely hot in the middle of the day through the summer months. Either of the borders would do, and the first sowing should come as early as the weather will permit and the soil is dry. Make shallow drills in the soil with a hoe and scatter the seed quite thickly. Cover it lightly and press the soil down firm. The rows can be as close as the width of your hoe. As soon as the plants are three or four inches high, pull out the larger plants and send to the cook. Two weeks later all can be gathered as fast as wanted. In my garden in 1888 the first planting was made April 16th and the first crop was gathered May 26th. Two crops of spinach can be taken off before it is time to set out the celery. A fall crop should be sown in any spare place that can be found about September 1st. This crop, too, can be planted quite thick and two gatherings made, one to thin out and the second to clear off what is left. Of course it may happen that more can be gathered each time than is wanted. The idea is simply to pick the spinach twice during its growth, at such intervals and in such quantities as may be needed. My fall crop in 1887 was planted September 6th, and was all consumed before the ground froze hard in November, being gathered in all six times, giving about a peck at each picking. In such a city-lot garden the whole of one of the side borders would not be too much space for fall spinach, two sowings being made, one about August 15th and the second September 1st.

Another good plan would be to plant in September, thin out the plants and then to cover them over with old hay or straw for the winter. The only objection is the difficulty of getting suitable material to cover the plants, and the litter it would make in a place that the house-mother would prefer to see kept exquisitely neat. You can't run a city lot like a market garden, and the best plan is to consume the fall spinach and not attempt to carry it over the winter. The best variety to use here is the Savoy-leaved Spinach.

In addition to these varieties of vegetables, rhubarb, cucumber, tomato, celery, lettuce and spinach, some space should be given to the small green crops, parsley, cress, onion sets, etc., described in the preceding chapter. These crops will pretty closely fill the three borders, particularly as a liberal quantity of celery and lettuce will be needed. One of the side borders entirely devoted to celery will not be too much, as a family of five can easily dispose of the six dozen heads it will contain. The crop can be stored in the cellar and kept for use through the early part of the winter. With a little care the spinach crop can be made to fill all gaps left by the removal of other crops. Judicious crowding and double cropping are essential in such a doyley garden as this.

CHAPTER IX.

A CITY FRUIT GARDEN.

THERE are many small home lots with excellent soil and a good sunny aspect, where the tenant or owner would gladly have a garden were there time to attend to it. This spring work of planting, this weeding, raking, re-planting and frequent harvesting demands more time than can be afforded. You are busy in town all day, and it is only once in a while that half a day can be spared for the garden. The 25x60 yard is there, but it must be laid down to grass, because that requires only one planting in several years, and the mowing need only take an hour or so twice a month. The grass is a cheap carpet on which to spread the clothes, or it is the children's play-ground, and it is not necessary to have it kept like a lawn.

Still, if it could be made to pay a return, it would help wonderfully in the little matter of living expenses. There are crops that would just meet your wants, crops that require only one planting in two or three years, and some that will even last half a life-time. There is asparagus and rhubarb, and perhaps some of the small fruits. In such a small garden it would not be well to plant all the small fruits, because some kinds require too much room and show an unruly spirit in the matter of running about the estate. The fruits for your purposes would be the strawberry, currant and grape. These would give something for the table from early summer till late in the fall, excepting for about three weeks in August and September. This gap might be filled by raspberries and blackberries, but these plants would be unsuitable for so small a place.

In order to utilize your space to the utmost advantage and to place your crops in the best aspect, they should be arranged in the following order: First (supposing your plot to lie north and south) comes the rhubarb at the north or warmest end, as it is the first thing to start in the spring and needs the benefit of the sun and shelter. Next, the asparagus, then the strawberries, and, lastly, at the south or shady end, the currants, as they will submit more gracefully to the shadows than any of the others. The grapes will extend along the fence on each side. The accompanying diagram shows how the different crops may be mapped out. At the northern end next the fence is a bed three feet wide the whole width of the lot. This will contain five plants of rhubarb. Two paths, each two feet wide and placed three feet from the side fences, give access to the other beds. Between the paths and next the rhubarb bed is a space 15x15 feet that may be set out as a permanent asparagus bed, the plants standing in rows three feet apart. Next to this is a space 15x25 feet that should be set out with strawberry plants, placed one foot apart each way, there being room for three hundred and fifty plants. South of these is room for a dozen currant bushes, in three rows of four each. At the sides are the grape borders, three feet wide the whole length of the lot, there being room for sixteen vines, eight on each side.

This arrangement of the lot will give the greatest space to the plants and the least trouble in caring for

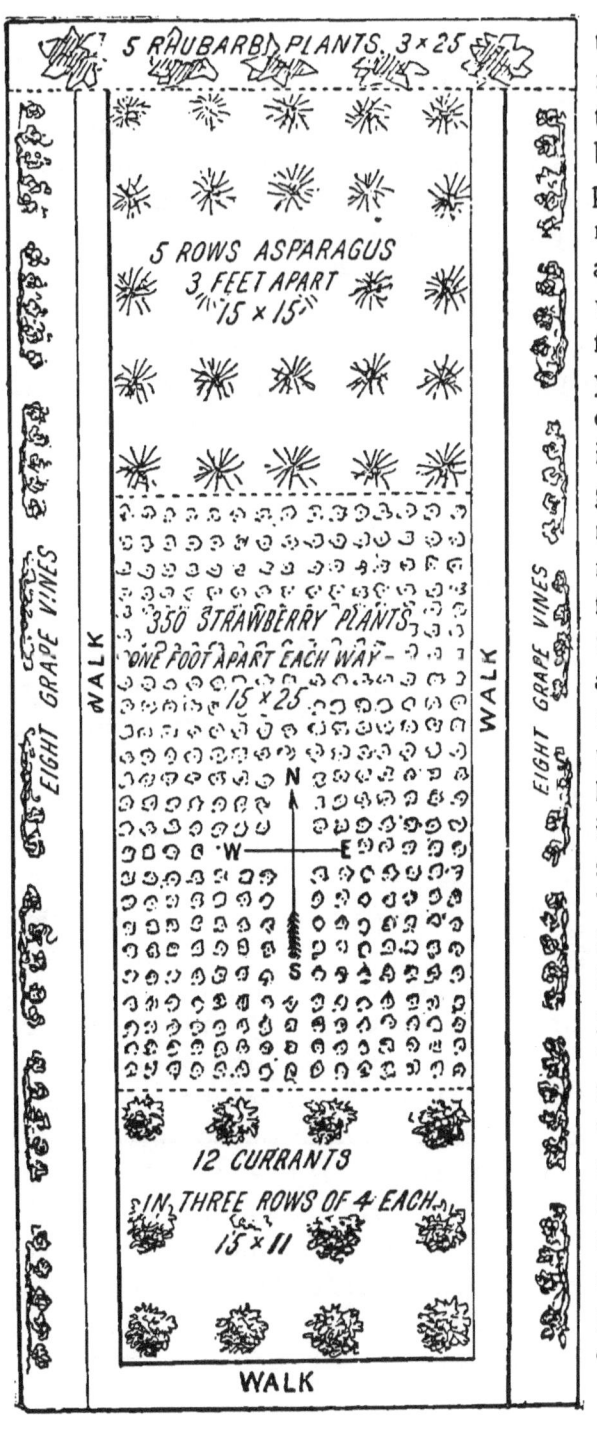

them, with ready access to all the beds. Once planted, the rhubarb and asparagus will remain for many years without replanting. The grapes will not require re-planting for at least ten years, and the currants will, with care, keep in bearing about six years. The strawberries will last two years, and with care three years, without renewal. By using these fruits we get rid of a large part of the constant re-planting of the vegetable garden, and the investment once made pays its little dividends year after year. Of course the first cost will be much greater than for seeds, but this is offset by the fact that the investment is made for years instead of weeks. At the regular catalogue rates the rhubarb will cost about $1.00; 75 asparagus plants, $1.00; 350 strawberries, $10.50; 12 currants, $1.50, and 16 grape vines, $8.00. This would be $22.00 for the whole. The labor of planting, fertilizers and tools would be the only additional expense, and once set out all expense would cease for at least two years, except the very slight expense of keeping the place in good order. From November 15th (in latitude of New York) to April 15th there would be absolutely no outlay whatever. On the other hand, we must forego all returns till the second year, and even then be content with only partial returns from a portion of the investment. On the third year everything would pay a dividend, and with the exception of the strawberries continue to pay every year for at least three years, and for the grapes, asparagus and rhubarb for from ten to twenty years to come. It costs more at first, yet in the long run the fruit garden pays as well, if not better, than the kitchen garden.

The care and culture of such a handkerchief fruit garden is simple and inexpensive, and when once planted consists chiefly in keeping the ground clean and open to the air. As in building a house, the first care is to get a secure foundation, so in our fruit garden the first consideration is the soil. It must be good and it must be dry. A clay soil, where the rain lingers on the surface after every shower, will never do, and if there are little pools to be seen for an hour or two after the rain has ceased to fall, the place must be drained, or it is better not to plant anything.

How your lot is to be drained depends on so many things peculiar to your soil and location, that you must trust wholly to yourself or some competent neighbor who is familiar with the lay or slope of your land and its immediate neighborhood. However, as nearly all our city lots are carved out of fields and orchards in suburbs and "additions," there is usually no need of drainage. Besides, water in the soil is quite as bad for you and your babies as for your plants, and it is not to be presumed you would consent to buy or build a house on wet land. As for the soil itself, if not good, it can usually be made good by carting in good soil from some field or pasture. Commonly, in newly made districts in the borders of our towns and cities, the soil is good enough for all practical purposes. All that it will probably need will be plenty of manure. Twenty-five dollars' worth will be little enough, and if you can afford to spend twice as much, it will not be too much. We are planting now once for all, and by a liberal expenditure now, money will be saved in the future. If you have possession of the place in the fall, by all means have the soil carefully spaded up and left rough through the winter. By so doing, the soil is exposed to the frost and air and broken up fine, and eggs of insects are destroyed. All the plants, except the strawberries, can be planted in the fall, but as there is a certain amount of risk from injury by ice, it is better to plant in the spring. The only advantage of fall planting is the time gained, as every one is in a hurry in the spring, and unless you give your orders early, the plants may arrive late or in bad condition.

Let us look at each crop in detail for a moment. The rhubarb will be the first thing to start in the spring, and the roots must be set out as early as possible after the frost leaves the ground. For this reason it is perhaps better, if just as convenient, to set this crop in the fall. The soil should be forked up and made free from stones, and heavily manured. If spread over the soil two or three inches deep, it will not be too much, as this plant has a tremendous appetite, and the bigger the dinner you give it, the bigger the rhubarb pies it will give you. The plants are mere clumps of fleshy roots and should be set at equal distances across the end of the lot, and just deep enough to have the top or crown covered about three inches. Spread straw or manure over the surface after the planting, and leave the bed till spring. If planted in the spring, the roots must be obtained very early and set in the ground with the deep red buds just under the surface. The after-culture the first year is to merely keep the weeds down, and there is no better way to do this than to rake the soil over lightly after every rain. When flower stalks appear, cut them off as soon as they rise above the leaves. In the fall spread more manure over the surface around the plants, as soon as the leaves die and disappear. The following spring the first crop of stalks can be gathered. Do not be too greedy. Give the plants a chance and gather only a few from each, rather than a quantity from one. Two stalks from each plant at a picking is enough, and once a week through the season is all you should expect. After that year perhaps twice as much can be taken every year for many years to come. Take good care of the plants, feed them well, and they will care for you and supply perhaps the best early crop you can give to the young folks round your table.

The asparagus bed needs a richer and deeper soil than any other crop. There is little danger of planting in too rich a soil, for you must remember that the bed is to be laid down once for all. and need not be re-planted for twenty years. Spring is the best time to plant and the earlier the better, for the asparagus is among the first to stir with life after the frosts have gone. The roots should be carefully planted in rows three feet apart, taking pains to spread them out carefully in a shallow trench, about a foot apart in the rows and with the point of the root buried about three inches. As soon as the shoots appear above ground, keep the soil light and open till the shoots begin to shade the ground with their tall feathery plumes. The weeds will then die out in the shade, and the last part of the season the bed will require very little attention. Do not cut any stalks, but let all grow as they will. Your object the first year is to get good strong plants, and you must wait till the following spring for the first dividends laid on the dinner table.

Strawberries are also good eaters, and it will pay well to treat them well. Your object is to obtain the greatest possible return from the smallest space, and to do this the best plan is to set the plants one foot apart each way, covering the entire space between the walks. As the plants are so near, you must not trouble them to search for food. Each plant must be so well fed that it is content to stay in its own limited spot of soil and not send its roots wandering off in search of something to eat. Neither must any plant be allowed to send out runners. In June and July, when runners appear, they must be rigorously cut off. If taken in hand early, when the runners are soft and green, they can be pulled off without trouble. A girl of ten ought to keep all the runners down, by going over the bed two or three times a week for about a month. A narrow rake is the best tool to keep the weeds down between the plants and to keep the soil loose after rains. When the ground freezes in December, the whole bed must be well covered with leaves or straw. The following spring it should be removed as soon as the frost is out of the ground, and then be put about the plants when the crop begins to ripen and again taken off after the crop is gathered.

The twelve currants should stand in three rows, about three feet apart and four in a row. The first year all that is needed is good rich soil and constant regular culture. In winter the plants take care of themselves, and all that is needed is to leave the soil loose and well broken up the last thing in the fall. The second year a small sample crop may be gathered just by way of encouragement. The third year the bushes will bear abundantly. The bushes are greatly benefitted by careful training. Cut out each spring all stems that cross other stems or that crowd each other, and all stems that show signs of decay or injury from any cause. The chief object is to have open, well shaped and rather short bushes with clean, healthy stems and free from suckers or adventurous shoots springing up round the roots. A single stem with a branching head is perhaps the best form.

The grapes should be planted in the spring as early as convenient, making good large holes in the soil to receive the roots, and spreading each root out carefully in its proper place. An entire book might be written on the training of grapes. There are as many ways as there are varieties, and it will be a good plan to read some good book on the grape and its culture before planting. The idea on which all systems of training the grape vine are based is very simple. The fruit is borne on green wood of the present season that grows out of ripe wood that grew the previous season. Your object must always be to have good ripe wood from which next year will come the green shoots bearing the crop. Your young plant will consist of a short stem with half a dozen buds (more or less) and a bunch of roots. In the spring nearly all of these buds will swell and send out tender young shoots. Wait till all are firmly started, and then with the fingers break off all but the largest and best nearest the ground. Do not use a knife, as the young plant may "bleed" or lose sap and be injured. Then carefully train this one shoot straight up the trellis or fence, tying it up as it grows and letting it grow as long as it will. About the middle of August pinch off the tip end of the growing shoot, and the green wood will slowly harden or "grow ripe" through the fall months. If side shoots start out from this stem, pinch the tips of each as soon as they appear to prevent them growing any longer. Better one good, stout shoot or cane, thickly covered with buds, than six poor, thin shoots with weak buds. After the leaves fall, this shoot should be cut down to about three feet from the ground. You now have a short, stout cane, from which will spring next year both fruit-bearing shoots and new canes for another year.

Having obtained a good cane with a dozen buds, any system of training may be followed that you fancy. Permanent canes may be trained along the bottom of the trellis, or spread over it in a fan-shape, or in any other way you please, provided always there is space between the canes for the new crop of wood that bears the fruit crop. In such a small fruit garden perhaps the best plan is to train up a single straight cane to the top of the fence, and to keep it there year after year. The bearing shoots will spread out three feet on each side, and the eight vines on the fence will cover all the wall space you have. The training of grape vines is an accomplishment well worth studying. It demands very little labor and gives an admirable chance to show that you are skillful at artistic effects. Nothing is more pliable than a grape vine, and with taste and a little patience your fence can be made to produce a beautiful effect, to say nothing of big crops. A neglected or ill-trained vine is simply ugly and unproductive. A well-trained vine is both a picture and a continued source of pleasure. Even in winter the well proportioned canes, neatly tied to the trellis, can be made quite effective as a bit of wall decoration.

This use of a city yard for a fruit garden costs more at first than if the lot is used for a vegetable garden, yet it pays quite as well, as the crops, when they do come, are worth more and last longer. The whole plantation is in the nature of a permanent investment in pleasure and profit. The rhubarb, asparagus and grapes will produce with care regular crops every season for many years, and even the currants will not require renewal more than once in ten years. The strawberry bed will be the least permanent, as it cannot be kept in good condition more than three or four years. On the other hand, it is easily renewed in one season and by a judicious system of re-planting not a single crop need be lost, though occasionally only half a crop will be gathered, while a portion of the bed is being renewed. The best way to do this will be to dig up and throw away half the plants as soon as the crop is gathered, and to re-plant the ground in August. These plants will give a crop the next year, when the rest of the old plants can be renewed in the same way. It is the same with the currants. As soon as the bushes begin to show signs of giving out, pull a part of them up and set new plants. It will be found a good plan to put a few cuttings in the ground each fall for new plants. In a year they will be well rooted, and can be used to take the place of any of the old plants that are worn out or injured by insects.

CHAPTER X.

THE CONCLUSION OF THE WHOLE MATTER.

OCTOBER days, and every garden finds its crop of facts, fruits and figures. My handkerchief garden in this year of grace, 1888, gave a variety of good things each in its season, all of which were eaten with a cheerful spirit. And herewith are the facts and figures. There were dry days and wet, total failures and big successes, weeds and bugs, lots of good, hard work, and altogether a fair return for the labor and money spent. Once or twice the crops overran the home market. The Early Jersey Wakefield and Early Summer cabbages ripened at about the same time, the first heads coming July 15th and the last being disposed of August 8th. A family of three can hardly master twelve heads ripe at one time, and as the industrious slug was always ready to lend a hand a large part of the crop was sacrificed on the altar of friendship. It was the same with cauliflowers. Eighteen Early Snowballs at one time was a little too much of a good thing, and the surplus was presented to appreciative neighbors.

Herein is one of the great advantages of a home lot—it tends to cultivate a generous heart in the gardener and a thankful spirit in his friends. You are pretty sure to have at times more than you want, and there is no better compliment to pay to a friend than a basket of fresh vegetables, wet with the dew, and right out of the home garden. People receive it with an expression that seems to say, "How very sweet in you, to be sure," and back comes the basket with a note of thanks calculated to fill the heart with the conviction that the world is not all a hollow show after all. There is one thing you can always present to a lady, and that's a flower. Why not a cauliflower? Is not a cabbage a green rose? Pull off the outer petals till the white heart begins to shine, pack it in a neat basket and send it to your friend's table. If she's a housemother with a soul above Kensington crewel she will say, and say truly, it is beautiful. Green lawns and shrubbery have their own glory, but there is also a glory of the cabbage patch and, though "the glory of one star" may not be "as the glory of another star," who shall say which is the greater glory?

My home lot account was opened November 1st, 1887. There were then on hand in the garden sixty-four plants of the Jessie strawberry, worth say 64 cents, a lot of currant cuttings worth $1, tools, flower pots and odd things worth $2, giving a stock on hand to begin the year of $2.64. There was paid out between that time and September 5th, 1888, just $14.64. Of this $4.40 was for 200 pounds of Mape's fertilizer, $1.10 for materials for a cold frame covered with protective cloth and $1.75 for labor. Of course, if this sum had been invested in vegetables it would have supplied the table for our small family for many weeks. Did it pay to spend it on the home lot?

The first return from the garden came on May 17th For the first ten days radishes were the only cheering thing to suggest the spring. Then the spinach began on May 26th, and soon came in a flood. By the first week in June young onions, beet tops and lettuce began to add a pleasant variety. On the 17th peas came also, and on the 23d strawberries. On the 24th the first delicate heads of young cabbage, and two days later the first potatoes. Beans and cauliflowers welcomed July, and then there was more than the home market could absorb. On the first day of August the total receipts from the garden, at the retail prices reported by THE AMERICAN GARDEN and in the village stores, amounted to just $23.10, or $8.13 in money value over the cash cost.

Through August the crops kept the table supplied with everything needed. Cucumbers came into bearing on the first of the month, and tomatoes soon after. On several days there were five kinds of vegetables served at one meal. There were no days without one, with two for an average. The crop of early potatoes was very small—a practical failure—and as there was no room for more, potatoes had to be purchased again by the last of August. As it was, one peck of choice potato seed supplied a crop that carried the house for six weeks, this being a decided failure. Turnips and beets did not do very well, except a small lot of beets on trial Among the best of the beets are the New Eclipse and the Early Dewing. The seed was from Philadelphia, and the beets proved of excellent quality. The Eclipse grew to enormous size, and were of fine flavor. Of onions a row of white silverskin onions proved to be of medium size, but of fine quality; name not known.

Cory corn was tried, but proved a total failure. Th late crops of peas were also complete failures Ot squashes I tried, at the request of THE AMERICAN GARDEN, the Sibley Winter squash. It proved to be a small smooth, pale green squash of superior flavor. The plants were enormous growers, but the crop was small and late. The Woodbury squash also proved to be a vigorous plant, but with me a poor bearer, one hill of four plants producing only three good squashes, resembling in appearance a Hubbard squash. A new red cabbage sent to me grew to a very great size, with round, compact heads. So much for a few experiments. A home lot always has its advantage— it is at once school, experimental station and a source of amusement. You never can tell how things will turn out. The beautiful pictures of the seed catalogues are even more splendid in reality at times, and then at times they lead to a high opinion of the lively imagination of the artists. Even with the losses it pays to try things, just for the sake of finding out for yourself. However, if you are looking for profit and not facts, don't do it.

After the first of June the home lot took very little labor or time. A good raking of the ground once or twice a week kept the soil in good order, and fifteen minutes or less every morning served to gather the crops. Up to September 1st the garden had produced crops valued at the retail price at $28.64. By the middle of September, $36.79. There were then on hand and unconsumed in the garden 200 good plants of celery, which, at 8 cents each, would be $16, and sundry other vegetables, including a large patch of spinach, about $2 more. Among other things produced were ninety good yearling currant bushes and about forty grape vines raised from cuttings. The currants would cost at least 10 cents each were I to buy them for the new garden I propose to plant next year, and this is a fair return from the garden of $9 The grape cuttings are also intended for the new place and will save at least 10 cents each, as they are all choice kinds, thus making a return of $4. These things are as much a crop as cauliflowers. They save buying plants next spring, and it is perfectly fair to add them to the returns from the garden. Had they not been wanted, of course the ground would have produced some other crop of less or equal value. The strawberry bed of sixty-four plants gave also 200 new plants, their value at 1 cent each being included in the return from the garden.

The sixty-four Jessie strawberry plants in my garden gave us, between June 21st and July 7th, just 23 quarts of very fine strawberries. The berries were uniformly large, some of the very largest berries being picked in the last quart. The place is too shaded for the best results, and I think with more sunshine and a trifle more rain they would have done much better. The flavor is "piney," bright and spicy. We bought no berries at the stores, as these were so fine. I propagated extensively for a new plantation for next year. I would decidedly recommend the Jessie for small gardens and as a rather late crop. My patch was entirely in hills, one foot apart each way, and carried a crop of lettuce between the plants early in the spring. Such close planting is a bit troublesome in gathering the crop, yet if you have only a home lot it must be done where profit is to be regarded.

As a whole my particular home lot was a happy one. Nearly everything bore fair crops, and at different times during the season my table was supplied with the following fruits and vegetables: Strawberries, radish, peas, spinach, onions, beans, lettuce, turnips, beets, carrots, tomatoes, cabbage, cauliflower, Upland cress, chicory, potatoes, squashes, cucumbers, parsnips and celery. From a housekeeping point of view really cheering; from a financial point of view quite as cheerful.

The grand total produced in the garden during the season was $54.79 for fruit and vegetables actually consumed. The season began with a stock on hand of tools, plants, cuttings, etc., of $2.64. It ends with a stock of ninety currants at ten cents, $9; forty grape vines at ten cents, $4; 200 strawberry plants valued at $2; tools, etc., $3; making a total of $18. This is real profit, and should be credited to the garden, for the new stock of plants helps to reduce the cost of the new garden to be planted next year. My home lot was a nursery as well as a garden, and returned a nurseryman's profits, and the whole of his profits, because the stock if bought must be paid for at retail prices.

Did the home lot pay? Was the return sufficient for the labor? It was, and the garden did pay, because the time spent on it was odd time not available for other work. Besides this, the work was a pleasure and a sanitary measure, paying a big dividend in red blood, sound sleep, a good appetite and a cheerful spirit. If you have ever been sick and paid doctor's bills you will know just what these things are worth in hard cash. The cost in money was $14.64 and about thirty days' labor between March and November. The entire return, including new stock valued at $15, was $69.79. Taking cash spent from this leaves $55.15, or about $1.80 for each day's labor spent in the garden. Of course if the labor had all been hired at the regular rate here of $2 the garden would have been carried on at a loss.

This brings the whole matter down to a business

basis, where you can settle for yourself whether it will pay you to have a home lot garden. Is your time worth so much (and it will have to be worth a good deal more than the average) that your unemployed minutes afternoons and before breakfast are worth more than 18 cents an hour? If they are a home lot will not pay you. If they are not, and if you consider health, fresh and superior vegetable food worth anything, then a home lot will pay you, as it did me, big dividends. For the great majority of families, particularly where there are young people who can help out-of-doors, a home lot will make just the difference between profit and loss, between money in the savings bank and unpaid bills at the stores. The home lot is the one reliable asset in your little property that will neither fail, fly away to Canada nor pass its dividends—the one partnership in which you will always hold a controlling interest.

Look at it in any way you will, keep a garden for pleasure or profit or health, you may set it down as your personal as well as national duty to make the most of the land that has been given to you. It is my belief that every man who has a bit of land is bound to consider it as a trust whereof he shall render account and wherewith he should do his best to make the earth bring forth her increase for the benefit of himself, his folks and the rest of the republic.

THE END.

www.ingramcontent.com/pod-product-compliance
Lightning Source LLC
Chambersburg PA
CBHW022152090426
42742CB00010B/1481